Guide to Green Fabrics Student Workbook™ | Chapter Lessons

Chapter Lessons | **Student Workbook**
Guide to Green Fabrics

© 2013 Guide to Green Fabrics Student Workbook™

All rights reserved. No part of this book may be reproduced or transmitted in any form or by any means, electronic or mechanical, including photocopying, recording, or by any information storage and retrieval system, without permission in writing from the publisher. Exceptions may be made for very brief excerpts used in published reviews.

This publication is designed to provide information only. It is sold with the understanding that the author and publisher are not engaged in rendering design, legal, accounting or other professional services. If legal or other expert assistance is required, the services of a competent professional should be sought.

The purpose of this manual is to educate and entertain. The author and publisher shall have neither liability nor responsibility to any person or entity with respect to any loss or damage caused, or alleged to have been caused, directly or indirectly, by the information contained in this book.

Published by:

Kristene Smith, Incorporated
Post Office Box 233553
Sacramento, California 95823

Library of Congress Control Number 2011909134

ISBN-13 9780989474733

First Edition

Guide to Green Fabrics Student Workbook™

Table of Contents

Lesson One: Abaca 3
Lesson Two: Bamboo 5
Lesson Three: Cervelt™ 7
Lesson Four: Coir 9
Lesson Five: Corn Husk 11
Lesson Six: Eco-Fi® 13
Lesson Seven: Eco-Leather/ E-Leather 15
Lesson Eight: Hemp 17
Lesson Nine: Henequen/ Sisal 20
Lesson Ten: Ingeo 22
Lesson Eleven: Jute 24
Lesson Twelve: Kenaf 26
Lesson Thirteen: Milk Protein 28
Lesson Fourteen: Modal 30
Lesson Fifteen: Natural Latex 32
Lesson Sixteen: Organic Cashmere 34
Lesson Seventeen: Organic Cotton 36
Lesson Eighteen: Organic Linen 38
Lesson Nineteen: Organic Wool 40
Lesson Twenty: Pina 42
Lesson Twenty-One: Ramie 44
Lesson Twenty-Two: Recycled Nylon 46
Lesson Twenty-Three: Recycled Polyester 48
Lesson Twenty-Four: Recycled Wool 50
Lesson Twenty-Five: Rush 52
Lesson Twenty-Six: Sea Cell 54
Lesson Twenty-Seven: Sea Grass 56
Lesson Twenty-Eight: Silk 58
Lesson Twenty-Nine: Soybean Protein 60
Lesson Thirty: Spider Silk 62
Lesson Thirty-One: Tencel®/ Lyocell 64
Lesson Thirty-Two: Wild Nettle 66

Chapter Lessons | **Student Workbook**
Guide to Green Fabrics

Chapter 1 | Abaca
Lesson One

Introduction – Three Things You Know
- Write down three things you know about abaca.
- Share your knowledge with the class and make a list on the board.
- You will have assigned reading and will watch a Power Point presentation. Afterward, write down three things you've learned from both of these exercises. Save your answers to share later.

Reading
- Read the abaca chapter.

Power Point Presentation
- Watch the abaca chapter Power Point presentation.
- Write down three things you've learned. Remember to save your answers to share later.

Mass Media
- Bring to class an article, print ad, online promotion, film review, or advertisement showing how green fabrics, green products, and related concerns are promoted or portrayed in the media. Break into groups to discuss how green fabrics and other environmental concerns are portrayed in the mass media.

Quick Answer: Sustainability Approach
- What would be your approach to developing an innovative solution to the following: consumer waste?

Three Things You've Learned
- Review your original abaca-related entries and compare them to what you've learned during the reading and Power Point exercises.

Abaca Chapter Quiz
- Take the abaca chapter quiz.

Abaca Chapter Quiz

Q1: In what way does abaca differ from true hemp? List the various differences between the two varieties.

_____

_____

_____

Q2: How has abaca impacted the economic fortunes of the Philippines?

_____

_____

_____

_____

_____

Q3: While the use of abaca in the fashion and furnishings industries is common, how does it aid people suffering from diseases like psoriasis and eczema?

_____

_____

_____

_____

_____

Q4: Grading of abaca fiber depends on certain production and drying techniques. List these.

_____

_____

_____

_____

_____

Chapter 2 | **Bamboo**
Lesson Two

Introduction – Three Things You Know
- o Write down three things you know about bamboo.
- o Share your knowledge with the class and make a list on the board.
- o You will have assigned reading and will watch a Power Point presentation. Afterward, write down three things you've learned from both of these exercises. Save your answers to share later.

Reading
- o Read the bamboo chapter.

Power Point Presentation
- o Watch the bamboo chapter Power Point presentation.
- o Write down three things you've learned. Remember to save your answers to share later.

Open Discussion
- o In terms of cultural exchange, education, and the passing of traditions, how can international governments work together to cross-train workers on the special skills needed to harvest and locally manufacture regionally grown sustainable fibers? Would this be viable? Does this pose problems relating to local economic development especially in developing nations? Remember to consider language barriers and *secrets of the trade* in your response.

Quick Answer: Sustainability Approach
- o What would be your approach to developing an innovative solution to the following: eco-messaging and advertising?

Three Things You've Learned
- o Review your original bamboo-related entries and compare them to what you've learned during the reading and Power Point exercises.

Bamboo Chapter Quiz
- o Take the bamboo chapter quiz.

Bamboo Chapter Quiz

Q1: Bamboo cultivation in China is extremely eco-friendly. Why?

_____

_____

_____

Q2: Clothes made of bamboo are also referred to as *air conditioning dress*. What unique properties make the dress cool and suitable for summer?

_____
_____
_____
_____

Q3: Bamboo viscose can mimic several other fibers and also does not use formaldehyde, which is a known carcinogen. Name the fibers it mimics and the quality of bamboo viscose which makes it an apt replacement for these fibers.

_____
_____
_____
_____

Q4: What quality of bamboo fiber makes it ideal for medical and health related products?

_____
_____
_____
_____

Chapter 3 | Cervelt™
Lesson Three

Introduction – Three Things You Know
- Write down three things you know about Cervelt™.
- Share your knowledge with the class and make a list on the board.
- You will have assigned reading and will watch a Power Point presentation. Afterward, write down three things you've learned from both of these exercises. Save your answers to share later.

Reading
- Read the Cervelt™ chapter.

Power Point Presentation
- Watch the Cervelt™ chapter Power Point presentation.
- Write down three things you've learned. Remember to save your answers to share later.

Celebrity Profile
- Using the internet and other resources, choose an eco-minded celebrity to profile in a short, yet complete, one-page report. Be sure to include the reasons why the celebrity is eco-minded and list some of their associations to the sustainability community.

Three Things You've Learned
- Review your original Cervelt™-related entries and compare them to what you've learned during the reading and Power Point exercises.

Cervelt Chapter Quiz
- Take the Cervelt™ chapter quiz.

Cervelt™ Chapter Quiz

Q1: Why is Cervelt™ considered an extremely expensive and rare eco-friendly fiber?

_____

_____

_____

_____

Q2: Cervelt™ remains pill free. What qualities of the fiber allow it to remain so?

_____

_____

_____

_____

Q3: Cervelt™ is made from raw material obtained from the red deer. Where is this breed of deer found?

_____

_____

_____

_____

Q4: What conditions are made available to the red deer to ensure its long life?

_____

_____

_____

_____

Chapter 4 | Coir
Lesson Four

Introduction – Three Things You Know
- Write down three things you know about coir.
- Share your knowledge with the class and make a list on the board.
- You will have assigned reading and will watch a Power Point presentation. Afterward, write down three things you've learned from both of these exercises. Save your answers to share later.

Reading
- Read the coir chapter.

Power Point Presentation
- Watch the coir chapter Power Point presentation.
- Write down three things you've learned. Remember to save your answers to share later.

Creative Worksheet
- Fashion Design. Sketch a green fashion design with the personality of a known celebrity. Submit a full color sketch or painting for this assignment.

Quick Answer: Sustainability Approach
- What would be your approach to developing an innovative solution to the following: natural dye and coloring?

Three Things You've Learned
- Review your original coir-related entries and compare them to what you've learned during the reading and Power Point exercises.

Coir Chapter Quiz
- Take the coir chapter quiz.

Coir Chapter Quiz

Q1: Elaborate on the connection between coir and India.

_____

_____

_____

_____

Q2: How tall can coconut trees grow?

Q3: How did colonialism help the spread of coir fiber and its usage?

Q4: Which well-known carpet firm used coir in fabric production?

Chapter 5 | Corn Husk
Lesson Five

Introduction – Three Things You Know
- Write down three things you know about corn husk.
- Share your knowledge with the class and make a list on the board.
- You will have assigned reading and will watch a Power Point presentation. Afterward, write down three things you've learned from both of these exercises. Save your answers to share later.

Reading
- Read the corn husk chapter.

Power Point Presentation
- Watch the corn husk chapter Power Point presentation.
- Write down three things you've learned. Remember to save your answers to share later.

Open Discussion
- How can the design industry improve the appeal of eco-products in general? With the sheer cost of green design being a factor, what messages and techniques can be employed using mass communications and consumer advertising?

Three Things You've Learned
- Review your original corn husk-related entries and compare them to what you've learned during the reading and Power Point exercises.

Corn Husk Chapter Quiz
- Take the corn husk chapter quiz.

Corn Husk Chapter Quiz

Q1: What are the similarities between corn husk fiber, cotton, and linen?

_____

_____

_____

_____

Q2: Name one of the most important products made using corn husk.

_____

_____

_____

_____

Q3: Why is corn husk fiber considered an extremely green option in spite of the chemicals used in producing the fiber?

_____

_____

_____

_____

Q4: Why is corn husk considered greener than traditional cotton?

_____

_____

_____

_____

Chapter 6 | Eco-Fi®
Lesson Six

Introduction – Three Things You Know
- o Write down three things you know about Eco-Fi®.
- o Share your knowledge with the class and make a list on the board.
- o You will have assigned reading and will watch a Power Point presentation. Afterward, write down three things you've learned from both of these exercises. Save your answers to share later.

Reading
- o Read the Eco-Fi® chapter.

Power Point Presentation
- o Watch the Eco-Fi® chapter Power Point presentation.
- o Write down three things you've learned. Remember to save your answers to share later.

Social/ Mobile
- o Explain how social media and mobile technology are empowering green designers to spread messages of environmental awareness across the globe. Elaborate its contribution in building a global community that supports green fabrics and makes efforts to promote it among peers and colleagues. Are these results better as compared to conventional media such as print and television? How so?

Quick Answer: Sustainability Approach
- o What would be your approach to developing an innovative solution to the following: marketing green fabrics?

Three Things You've Learned
- o Review your original Eco-Fi®-related entries and compare them to what you've learned during the reading and Power Point exercises.

Eco-Fi® Chapter Quiz
- o Take the Eco-Fi® chapter quiz.

Eco-Fi® Chapter Quiz

Q1: Compared to natural fibers, what are two of the advantages of Eco-Fi®?

_____

_____

_____

Q2: Enumerate some of the benefits of using Eco-Fi® relative to its care.

_____

_____

_____

_____

_____

Q3: Which company won the first UN Fashion Industry Award for Environmental Excellence?

_____

_____

_____

_____

_____

Q4: What qualities of Eco-Fi® make it a fabric perfect for use in outdoor sports and other allied activities?

_____

_____

_____

_____

_____

Chapter 7 | Eco-Leather/ E-Leather
Lesson Seven

Introduction – Three Things You Know
- o Write down three things you know about eco-leather/ e-leather.
- o Share your knowledge with the class and make a list on the board.
- o You will have assigned reading and will watch a Power Point presentation. Afterward, write down three things you've learned from both of these exercises. Save your answers to share later.

Reading
- o Read the eco-leather/ e-leather chapter.

Power Point Presentation
- o Watch the eco-leather/ e-leather chapter Power Point presentation.
- o Write down three things you've learned. Remember to save your answers to share later.

My Own Way
- o Have you ever been in a situation whereby you were confronted with challenges to your eco-friendly beliefs? What did you do, and how did others react?

Quick Answer: Sustainability Approach
- o What would be your approach to developing an innovative solution to the following: organic farming methods?

Three Things You've Learned
- o Review your original eco-leather/ e-leather-related entries and compare them to what you've learned during the reading and Power Point exercises.

Eco-Leather/ E-Leather Chapter Quiz
- o Take the eco-leather/ e-leather chapter quiz.

Eco-Leather/ E-Leather Chapter Quiz

Q1: What is the difference between conventionally processed leather and eco-leather?

_____

_____

_____

_____

Q2: What makes South America one of the largest providers of raw material for eco-leather?

_____

_____

_____

_____

Q3: What is a *growth mark* and how does it help in establishing the identity of eco-leather?

_____

_____

_____

_____

Q4: What happens to growth marks over time?

_____

_____

_____

_____

Chapter 8 | Hemp
Lesson Eight

Introduction – Three Things You Know
- Write down three things you know about hemp.
- Share your knowledge with the class and make a list on the board.
- You will have assigned reading and will watch a Power Point presentation. Afterward, write down three things you've learned from both of these exercises. Save your answers to share later.

Reading
- Read the hemp chapter.

Power Point Presentation
- Watch the hemp chapter Power Point presentation.
- Write down three things you've learned. Remember to save your answers to share later.

News Article Review
- Research a current news article on eco-friendly fabrics. Describe its contents in detail, and provide insights to the class about its relevancy to sustainability.

Quick Answer: Sustainability Approach
- What would be your approach to developing an innovative solution to the following: corporate subsidies for eco-farmers?

Three Things You've Learned
- Review your original hemp-related entries and compare them to what you've learned during the reading and Power Point exercises.

Hemp Chapter Quiz
Take the hemp chapter quiz.

Hemp Chapter Quiz

Q1: Hemp has a negative connotation because of its association with which plant?

_____

_____

_____

_____

Q2: In what way is hemp associated with the Declaration of Independence and the Founding Fathers of the United States?

_____

_____

_____

_____

_____

Q3: Elaborate on the connection between the U.S. military and hemp.

_____

_____

_____

_____

_____

Q4: Describe the climatic and soil conditions ideal for cultivating hemp.

_____

_____

_____

_____

_____

Q5: Hemp is a bast fiber. What is a bast fiber?

_____

_____

_____

_____

Q6: What is *retting* and what is the difference between traditional and modern retting methods?

_____

_____

_____

_____

Chapter 9 | Henequen/ Sisal
Lesson Nine

Introduction – Three Things You Know
- o Write down three things you know about henequen/ sisal.
- o Share your knowledge with the class and make a list on the board.
- o You will have assigned reading and will watch a Power Point presentation. Afterward, write down three things you've learned from both of these exercises. Save your answers to share later.

Reading
- o Read the henequen/ sisal chapter.

Power Point Presentation
- o Watch the henequen/ sisal chapter Power Point presentation.
- o Write down three things you've learned. Remember to save your answers to share later.

Essay
- o Write an 800-word essay on the impact of green design on the media. How prevalent is green design, and are there enough media worthy topics emanating from the industry to generate coverage? Which topics generate the most interest? Remember to use cited sources in your response.

Quick Answer: Sustainability Approach
- o What would be your approach to developing an innovative solution to the following: organic pest control?

Three Things You've Learned
- o Review your original henequen/ sisal-related entries and compare them to what you've learned during the reading and Power Point exercises.

Henequen/ Sisal Chapter Quiz
- o Take the henequen/ sisal chapter quiz.

Henequen/ Sisal Chapter Quiz

Q1: How is the economy of Mexico tied with sisal or henequen production?

_____

_____

_____

Q2: Elaborate on the link between the invention of the reaper and the use of sisal ropes.

_____

_____

_____

_____

Q3: What feature of the sisal plant makes it a favored crop in developing and third world nations?

_____

_____

_____

_____

Q4: The use of a certain chemical makes sisal fiber fire retardant. Name the chemical.

_____

_____

_____

_____

Chapter 10 | Ingeo
Lesson Ten

Introduction – Three Things You Know
- Write down three things you know about ingeo.
- Share your knowledge with the class and make a list on the board.
- You will have assigned reading and will watch a Power Point presentation. Afterward, write down three things you've learned from both of these exercises. Save your answers to share later.

Reading
- Read the ingeo chapter.

Power Point Presentation
- Watch the ingeo chapter Power Point presentation.
- Write down three things you've learned. Remember to save your answers to share later.

Celebrity Profile
- Using the internet and other resources, choose an eco-minded celebrity to profile in a short, yet complete, one-page report. Be sure to include the reasons why the celebrity is eco-minded and list some of their associations to the sustainability community.

Three Things You've Learned
- Review your original ingeo-related entries and compare them to what you've learned during the reading and Power Point exercises.

Ingeo Chapter Quiz
- Take the ingeo chapter quiz.

Ingeo Chapter Quiz

Q1: What does the term *Ingeo* mean?

_____

_____

_____

_____

Q2: Name some of the qualities that ingeo shares with synthetic fibers.

___

___

___

___

___

___

Q3: Which countries are the main producers of ingeo?

___

___

___

___

___

___

Q4: Describe the method used for producing ingeo fiber. What is ingeo used for?

___

___

___

___

___

___

Chapter 11 | Jute
Lesson Eleven

Introduction – Three Things You Know
- Write down three things you know about jute.
- Share your knowledge with the class and make a list on the board.
- You will have assigned reading and will watch a Power Point presentation. Afterward, write down three things you've learned from both of these exercises. Save your answers to share later.

Reading
- Read the jute chapter.

Power Point Presentation
- Watch the jute chapter Power Point presentation.
- Write down three things you've learned. Remember to save your answers to share later.

Mass Media
- Bring to class an article, print ad, online promotion, film review, or advertisement showing how green fabrics, green products, and related concerns are promoted or portrayed in the media. Break into groups to discuss how green fabrics and other environmental concerns are portrayed in the mass media.

Quick Answer: Sustainability Approach
- What would be your approach to developing an innovative solution to the following: green product packaging design?

Three Things You've Learned
- Review your original jute-related entries and compare them to what you've learned during the reading and Power Point exercises.

Jute Chapter Quiz
- Take the jute chapter quiz.

Jute Chapter Quiz

Q1: Apart from its color, why is jute called the *golden fiber*? Remember to consider jute's many levels of uses and appeal in your answer.

_____

_____

_____

Q2: Which natural substance found in jute aids in the attachment of short jute fibers to form longer ones?

Q3: Why is processing jute without chemicals a best practice, thus making it extremely environmentally-friendly?

Q4: During the retting process what action takes place in the plant's structure?

Chapter 12 | Kenaf
Lesson Twelve

Introduction – Three Things You Know
- Write down three things you know about kenaf.
- Share your knowledge with the class and make a list on the board.
- You will have assigned reading and will watch a Power Point presentation. Afterward, write down three things you've learned from both of these exercises. Save your answers to share later.

Reading
- Read the kenaf chapter.

Power Point Presentation
- Watch the kenaf chapter Power Point presentation.
- Write down three things you've learned. Remember to save your answers to share later.

Midterm Assignment Overview/ Sign-Up

Three Things You've Learned
- Review your original kenaf-related entries and compare them to what you've learned during the reading and Power Point exercises.

Kenaf Chapter Quiz
- Take the kenaf chapter quiz.

Kenaf Chapter Quiz

Q1: What other names are used for kenaf?

___

___

___

___

Q2: What is kenaf's native country and in what other countries is it now produced?

_____

_____

_____

_____

Q3: What are the two methods commonly used to extract fiber from the kenaf plant? Provide detail in your response.

_____

_____

_____

_____

Q4: Which country has labeled newsprint made from kenaf as *tree-free*?

_____

_____

_____

_____

Chapter 13 | Milk Protein
Lesson Thirteen

Introduction – Three Things You Know
- o Write down three things you know about milk protein.
- o Share your knowledge with the class and make a list on the board.
- o You will have assigned reading and will watch a Power Point presentation. Afterward, write down three things you've learned from both of these exercises. Save your answers to share later.

Reading
- o Read the milk protein chapter.

Power Point Presentation
- o Watch the milk protein chapter Power Point presentation.
- o Write down three things you've learned. Remember to save your answers to share later.

Midterm Worktime
- o Use this time to work on your midterm project. Remember to keep reading and researching at home as well.

Quick Answer: Sustainability Approach
- o What would be your approach to developing an innovative solution to the following: student eco-education?

Three Things You've Learned
- o Review your original milk protein-related entries and compare them to what you've learned during the reading and Power Point exercises.

Milk Protein Chapter Quiz
- o Take the milk protein chapter quiz.

Milk Protein Chapter Quiz

Q1: Name the ancient beauty associated with the use of milk as a beauty aid. Name one of her favorite uses for milk.

_____

_____

_____

Q2: During which decade was milk protein fabric invented, and where? Give some of the registered names for fabrics made from milk protein.

Q3: What fibers are commonly blended with milk protein fibers?

Q4: Explain *dewatering*.

Chapter 14 | Modal
Lesson Fourteen

Introduction – Three Things You Know
- o Write down three things you know about modal.
- o Share your knowledge with the class and make a list on the board.
- o You will have assigned reading and will watch a Power Point presentation. Afterward, write down three things you've learned from both of these exercises. Save your answers to share later.

Reading
- o Read the modal chapter.

Power Point Presentation
- o Watch the modal chapter Power Point presentation.
- o Write down three things you've learned. Remember to save your answers to share later.

Midterm Worktime
- o Use this time to work on your midterm project. Remember to keep reading and researching at home as well.

Quick Answer: Sustainability Approach
- o What would be your approach to developing an innovative solution to the following: recycling efforts?

Three Things You've Learned
- o Review your original modal-related entries and compare them to what you've learned during the reading and Power Point exercises.

Modal Chapter Quiz
- o Take the modal chapter quiz.

Modal Chapter Quiz

Q1: What is the difference between rayon and modal fabric?

_____

_____

_____

Q2: What is the formal, registered name for modal?

Q3: What negative aspect of natural fibers is eliminated in modal fabric?

Q4: What does *hygroscopic* mean, and how is modal hygroscopic?

Chapter 15 | Natural Latex
Lesson Fifteen

Introduction – Three Things You Know
- Write down three things you know about natural latex.
- Share your knowledge with the class and make a list on the board.
- You will have assigned reading and will watch a Power Point presentation. Afterward, write down three things you've learned from both of these exercises. Save your answers to share later.

Reading
- Read the natural latex chapter.

Power Point Presentation
- Watch the natural latex chapter Power Point presentation.
- Write down three things you've learned. Remember to save your answers to share later.

Midterm Presentations
- You will present your research project to the class. Presentations are 10 minutes each and must be clearly presented both orally and visually. Be prepared to receive constructive feedback and individual and group evaluations.

Three Things You've Learned
- Review your original natural latex-related entries and compare them to what you've learned during the reading and Power Point exercises.

Natural Latex Chapter Quiz
- Take the natural latex chapter quiz.

Natural Latex Chapter Quiz

Q1: What elements are found in latex which gives it its unique stickiness?

_____

_____

_____

_____

Q2: Describe the journey of the rubber tree from Brazil to Southeast Asia.

Q3: What is *bud grafting*?

Q4: What is the *cambium* and why is it important to ensure that the cambium remains undamaged during the tapping process?

Chapter 16 | Organic Cashmere
Lesson Sixteen

Introduction – Three Things You Know
- Write down three things you know about organic cashmere.
- Share your knowledge with the class and make a list on the board.
- You will have assigned reading and will watch a Power Point presentation. Afterward, write down three things you've learned from both of these exercises. Save your answers to share later.

Reading
- Read the organic cashmere chapter.

Power Point Presentation
- Watch the organic cashmere chapter Power Point presentation.
- Write down three things you've learned. Remember to save your answers to share later.

Midterm Presentations
- You will present your research project to the class. Presentations are 10 minutes each and must be clearly presented both orally and visually. Be prepared to receive constructive feedback and individual and group evaluations.

Quick Answer: Sustainability Approach
- What would be your approach to developing an innovative solution to the following: waterways and water supplies for organic farming?

Three Things You've Learned
- Review your original organic cashmere-related entries and compare them to what you've learned during the reading and Power Point exercises.

Organic Cashmere Chapter Quiz
- Take the organic cashmere chapter quiz.

Organic Cashmere Chapter Quiz

Q1: From where did the word *cashmere* originate?

_____

_____

_____

Q2: How and when did cashmere find its way to Europe?

_____

_____

_____

_____

_____

Q3: To whom is systematic cashmere production credited?

_____

_____

_____

_____

_____

Q4: How does the cashmere goat's *moulting* lead to the production of cashmere fiber? What is *dehairing*?

_____

_____

_____

_____

_____

Chapter 17 | Organic Cotton
Lesson Seventeen

Introduction – Three Things You Know
- Write down three things you know about organic cotton.
- Share your knowledge with the class and make a list on the board.
- You will have assigned reading and will watch a Power Point presentation. Afterward, write down three things you've learned from both of these exercises. Save your answers to share later.

Reading
- Read the organic cotton chapter.

Power Point Presentation
- Watch the organic cotton chapter Power Point presentation.
- Write down three things you've learned. Remember to save your answers to share later.

Creative Worksheet
- Interior Design. Develop a perspective (drawing) of a green interior design (single room only, any room choice) with the personality of a known consumer brand. Submit a full color drawing or painting only.

Quick Answer: Sustainability Approach
- What would be your approach to developing an innovative solution to the following: consumer eco-education?

Three Things You've Learned
- Review your original organic cotton-related entries and compare them to what you've learned during the reading and Power Point exercises.

Organic Cotton Chapter Quiz
- Take the organic cotton chapter quiz.

Organic Cotton Chapter Quiz

Q1: What is the difference between traditional and organic cotton?

_____

_____

_____

Q2: Which natural element of organic cotton is retained during processing lending to its elegant drape and fall?

Q3: What precautions are taken in order to ensure the environmentally positive attributes of organic cotton at the planting and harvesting stages?

Q4: What factors led farmers to discontinue the practice of crop rotation?

Chapter 18 | Organic Linen
Lesson Eighteen

Introduction – Three Things You Know
- Write down three things you know about organic linen.
- Share your knowledge with the class and make a list on the board.
- You will have assigned reading and will watch a Power Point presentation. Afterward, write down three things you've learned from both of these exercises. Save your answers to share later.

Reading
- Read the organic linen chapter.

Power Point Presentation
- Watch the organic linen chapter Power Point presentation.
- Write down three things you've learned. Remember to save your answers to share later.

Open Discussion
- In order to propel early, eco-friendly education, should public and other primary education systems develop mandated, environmentally-focused, educational outreach programs for students? What benefits would this provide? What kinds of opposition to this proposal might occur, and why?

Quick Answer: Sustainability Approach
- What would be your approach to developing an innovative solution to the following: government involvement and support of eco-fiber production?

Three Things You've Learned
- Review your original organic linen-related entries and compare them to what you've learned during the reading and Power Point exercises.

Organic Linen Chapter Quiz
- Take the organic linen chapter quiz.

Organic Linen Chapter Quiz

Q1: Which ancient civilization is closely connected with linen fabric?

_____

_____

_____

Q2: Name the countries that are the largest producers of linen fiber.

_____

_____

_____

_____

Q3: In what respect does linen score over its close competitor: cotton?

_____

_____

_____

_____

Q4: Linen is used to make space suits because of a certain characteristic unique to the fiber. Name the characteristic.

_____

_____

_____

_____

Chapter 19 | Organic Wool
Lesson Nineteen

Introduction – Three Things You Know
- Write down three things you know about organic wool.
- Share your knowledge with the class and make a list on the board.
- You will have assigned reading and will watch a Power Point presentation. Afterward, write down three things you've learned from both of these exercises. Save your answers to share later.

Reading
- Read the organic wool chapter.

Power Point Presentation
- Watch the organic wool chapter Power Point presentation.
- Write down three things you've learned. Remember to save your answers to share later.

Celebrity Profile
- Using the internet and other resources, choose an eco-minded celebrity to profile in a short, yet complete, one-page report. Be sure to include the reasons why the celebrity is eco-minded and list some of their associations to the sustainability community.

Quick Answer: Sustainability Approach
- What would be your approach to developing an innovative solution to the following: organic livestock?

Three Things You've Learned
- Review your original organic wool-related entries and compare them to what you've learned during the reading and Power Point exercises.

Organic Wool Chapter Quiz
- Take the organic wool chapter quiz.

Organic Wool Chapter Quiz

Q1: How does organic wool differ from conventional wool?

___

___

___

Q2: Which countries are the main producers of organic wool?

_____

_____

_____

_____

_____

Q3: What unique qualities of organic wool are responsible for its strength and durability?

_____

_____

_____

_____

_____

Q4: What prohibitive aspects of sheep rearing ensure the eco-quality of organic wool?

_____

_____

_____

_____

_____

Chapter 20 | Piña
Lesson Twenty

Introduction – Three Things You Know
- Write down three things you know about piña.
- Share your knowledge with the class and make a list on the board.
- You will have assigned reading and will watch a Power Point presentation. Afterward, write down three things you've learned from both of these exercises. Save your answers to share later.

Reading
- Read the piña chapter.

Power Point Presentation
- Watch the piña chapter Power Point presentation.
- Write down three things you've learned. Remember to save your answers to share later.

Social/ Mobile
- Discuss the role of social media and mobile technology in engaging the sustainability community and spreading awareness among the young generation about green fabrics and green design. How is it helping to create an informed generation that understands the values and benefits of green fabrics?

Quick Answer: Sustainability Approach
- What would be your approach to developing an innovative solution to the following: organic farming equipment?

Three Things You've Learned
- Review your original piña-related entries and compare them to what you've learned during the reading and Power Point exercises.

Piña Chapter Quiz
- Take the piña chapter quiz.

Piña Chapter Quiz

Q1: Name the plant from which piña fiber is extracted.

_____

_____

_____

Q2: Which countries are primarily associated with the cultivation of pineapples?

_____

_____

_____

_____

Q3: What other environmentally-friendly plants are closely associated with the production techniques of piña?

_____

_____

_____

_____

Q4: How is filament yarn created with piña and why does it lead to the presence of knots in the finished fabric?

_____

_____

_____

_____

Chapter 21 | Ramie
Lesson Twenty-One

Introduction – Three Things You Know
- Write down three things you know about ramie.
- Share your knowledge with the class and make a list on the board.
- You will have assigned reading and will watch a Power Point presentation. Afterward, write down three things you've learned from both of these exercises. Save your answers to share later.

Reading
- Read the ramie chapter.

Power Point Presentation
- Watch the ramie chapter Power Point presentation.
- Write down three things you've learned. Remember to save your answers to share later.

News Article Review
- Research a current news article on eco-friendly fabrics. Describe its contents in detail, and provide insights to the class about its relevancy to sustainability.

Quick Answer: Sustainability Approach
- What would be your approach to developing an innovative solution to the following: non-mechanical spinning and weaving methods?

Three Things You've Learned
- Review your original ramie-related entries and compare them to what you've learned during the reading and Power Point exercises.

Ramie Chapter Quiz
- Take the ramie chapter quiz.

Ramie Chapter Quiz

Q1: Elaborate on the affection shown toward ramie by the Chinese. Who was presented with gifts of ramie as a sign of respect and gratitude?

_____

_____

_____

Q2: Which special properties of ramie made it ideal as a wrapping for mummies?

Q3: What makes ramie fabric more susceptible to brittleness compared to other natural fibers? How does it rate in terms of resilience, elasticity, and elongation?

Q4: At what stage of the plant's growth is ramie harvested? Where is it best to cut the stem?

Chapter 22 | Recycled Nylon
Lesson Twenty-Two

Introduction – Three Things You Know
- Write down three things you know about recycled nylon.
- Share your knowledge with the class and make a list on the board.
- You will have assigned reading and will watch a Power Point presentation. Afterward, write down three things you've learned from both of these exercises. Save your answers to share later.

Reading
- Read the recycled nylon chapter.

Power Point Presentation
- Watch the recycled nylon chapter Power Point presentation.
- Write down three things you've learned. Remember to save your answers to share later.

My Own Way
- Have you ever been in a situation whereby you were confronted with challenges to your eco-friendly beliefs? What did you do and how did others react?

Quick Answer: Sustainability Approach
- What would be your approach to developing an innovative solution to the following: local farming and economic development?

Three Things You've Learned
- Review your original recycled nylon-related entries and compare them to what you've learned during the reading and Power Point exercises.

Recycled Nylon Chapter Quiz
- Take the recycled nylon chapter quiz.

Recycled Nylon Chapter Quiz

Q1: What is pre-consumer waste and how is it associated with recycled nylon fibers?

_____

_____

_____

_____

Q2: Name one product that uses recycled nylon fibers for production with the product itself being recycled for producing new recycled nylon fiber.

_____

_____

_____

_____

Q3: Sources must be accessed for raw material needed for recycled nylon production. Which two nations are the most significant for this purpose?

_____

_____

_____

_____

Q4: Describe the various steps, from accessing raw material to the finished fiber, that are used in producing recycled nylon.

_____

_____

_____

_____

Chapter 23 | Recycled Polyester
Lesson Twenty-Three

Introduction – Three Things You Know
- Write down three things you know about recycled polyester.
- Share your knowledge with the class and make a list on the board.
- You will have assigned reading and will watch a Power Point presentation. Afterward, write down three things you've learned from both of these exercises. Save your answers to share later.

Reading
- Read the recycled polyester chapter.

Power Point Presentation
- Watch the recycled polyester chapter Power Point presentation.
- Write down three things you've learned. Remember to save your answers to share later.

Essay
- Write an 800-word essay on the collective excitement presently seen in the design industry as indicated by the number of eco-fashion and interiors conferences and expositions being held on an international scale. Which kinds of attendees do these conferences attract? What features, speakers, and workshops are being presented to solidify the idea of green design?

Three Things You've Learned
- Review your original recycled polyester-related entries and compare them to what you've learned during the reading and Power Point exercises.

Recycled Polyester Chapter Quiz
- Take the recycled polyester chapter quiz.

Recycled Polyester Chapter Quiz

Q1: Name some of the most important qualities of synthetic fibers seen in garments made from recycled polyester.

_____

_____

_____

_____

Q2: What material is used to create recycled polyester?

Q3: What is *antimony* and how does it harm the environment?

Q4: How many barrels of oil are used annually to produce original virgin polyester?

Chapter 24 | Recycled Wool
Lesson Twenty-Four

Introduction – Three Things You Know
- Write down three things you know about recycled wool.
- Share your knowledge with the class and make a list on the board.
- You will have assigned reading and will watch a Power Point presentation. Afterward, write down three things you've learned from both of these exercises. Save your answers to share later.

Reading
- Read the recycled wool chapter.

Power Point Presentation
- Watch the recycled wool chapter Power Point presentation.
- Write down three things you've learned. Remember to save your answers to share later.

Open Discussion
- Describe your version of ideal government support systems for green designers. What would this look like and what incentives might there be? Is there a role for the government in green design?

Quick Answer: Sustainability Approach
- What would be your approach to developing an innovative solution to the following: environmentally-friendly shipping methods?

Three Things You've Learned
- Review your original recycled wool-related entries and compare them to what you've learned during the reading and Power Point exercises.

Recycled Wool Chapter Quiz
- Take the recycled wool chapter quiz.

Recycled Wool Chapter Quiz

Q1: How is wool produced?

_____

_____

_____

Q2: What is the difference between the hairy fiber of sheep and fibers obtained from other animals?

Q3: Describe *crimp* and how it aids in spinning.

Q4: What quality of wool makes it the favorite fiber of the Bedouins?

Chapter 25 | Rush
Lesson Twenty-Five

Introduction – Three Things You Know
- Write down three things you know about rush.
- Share your knowledge with the class and make a list on the board.
- You will have assigned reading and will watch a Power Point presentation. Afterward, write down three things you've learned from both of these exercises. Save your answers to share later.

Reading
- Read the rush chapter.

Power Point Presentation
- Watch the rush chapter Power Point presentation.
- Write down three things you've learned. Remember to save your answers to share later.

Social/ Mobile
- Describe how social media and mobile technology can assist up and coming green designers with their business practices. Does social media allow for instant promotion of eco-brands, and does the public respond to these messages? How can mobile technology help with the organization of one's green design business?

Quick Answer: Sustainability Approach
- What would be your approach to developing an innovative solution to the following: sustainable harvesting methods?

Three Things You've Learned
- Review your original rush-related entries and compare them to what you've learned during the reading and Power Point exercises.

Rush Chapter Quiz
- Take the rush chapter quiz.

Rush Chapter Quiz

Q1: What is *tatami*?

_____

_____

Q2: Why are items made of rush extremely lightweight?

Q3: When is the best time to gather rushes for the purpose of making fibers?

Q4: Why is it important to ensure that harvested rush plans are kept in the dark and kept damp?

Chapter 26 | Sea Cell
Lesson Twenty-Six

Introduction – Three Things You Know
- Write down three things you know about sea cell.
- Share your knowledge with the class and make a list on the board.
- You will have assigned reading and will watch a Power Point presentation. Afterward, write down three things you've learned from both of these exercises. Save your answers to share later.

Reading
- Read the sea cell chapter.

Power Point Presentation
- Watch the sea cell chapter Power Point presentation.
- Write down three things you've learned. Remember to save your answers to share later.

Creative Worksheet
- Textiles. Research and present a short oral report on the use of a specific green fiber providing detail on its origins, properties, and characteristics. Explain its weave, thickness, and durability, and discuss its availability in the marketplace.

Three Things You've Learned
- Review your original sea cell-related entries and compare them to what you've learned during the reading and Power Point exercises.

Sea Cell Chapter Quiz
- Take the sea cell chapter quiz.

Sea Cell Chapter Quiz

Q1: What is the raw material for making sea cell fabric and what process is used in its manufacture?

_____

_____

_____

_____

Q2: How does the body absorb nutrients released by sea cell fabric?

_____

_____

_____

_____

Q3: Which garments are most often created using sea cell fibers?

_____

_____

_____

_____

Q4: How is seaweed used by different cultures?

_____

_____

_____

_____

Chapter 27 | Sea Grass
Lesson Twenty-Seven

Introduction – Three Things You Know
- Write down three things you know about sea grass.
- Share your knowledge with the class and make a list on the board.
- You will have assigned reading and will watch a Power Point presentation. Afterward, write down three things you've learned from both of these exercises. Save your answers to share later.

Reading
- Read the sea grass chapter.

Power Point Presentation
- Watch the sea grass chapter Power Point presentation.
- Write down three things you've learned. Remember to save your answers to share later.

Mass Media
- Bring to class an article, print ad, online promotion, film review, or advertisement showing how green fabrics, green products, and related concerns are promoted or portrayed in the media. Break into groups to discuss how green fabrics and other environmental concerns are portrayed in the mass media.

Three Things You've Learned
- Review your original sea grass-related entries and compare them to what you've learned during the reading and Power Point exercises.

Sea Grass Chapter Quiz
- Take the sea grass chapter quiz.

Sea Grass Chapter Quiz

Q1: What is sea grass and where is it found?

_____

_____

_____

_____

Q2: Which countries are the chief cultivators of sea grass?

Q3: In what ways is sea grass an environment protector apart from its use as a green fiber?

Q4: How does sea grass get its unique look?

Chapter 28 | Silk
Lesson Twenty-Eight

Introduction – Three Things You Know
- o Write down three things you know about silk.
- o Share your knowledge with the class and make a list on the board.
- o You will have assigned reading and will watch a Power Point presentation. Afterward, write down three things you've learned from both of these exercises. Save your answers to share later.

Reading
- o Read the silk chapter.

Power Point Presentation
- o Watch the silk chapter Power Point presentation.
- o Write down three things you've learned. Remember to save your answers to share later.

Essay
- o Write an 800-word essay on what it would take for a designer to redirect their career from being a traditional designer to a green designer. What kinds of societal or emotional occurrences might take place that would cause such a thing to happen? What kind of industry exposure and education would they need to have in order to be effective in this new career choice? Would it be important for their personal values to be promoted through their brand? Explain your insights in detail.

Three Things You've Learned
- o Review your original silk-related entries and compare them to what you've learned during the reading and Power Point exercises.

Silk Chapter Quiz
- o Take the silk chapter quiz.

Silk Chapter Quiz

Q1: Explain the connection between silk and tea.

_____

_____

_____

_____

Q2: What is the Silk Route?

_____

_____

_____

_____

_____

Q3: Name some of the historical personalities that showed affinity for silk fabrics.

_____

_____

_____

_____

_____

Q4: How did silk making reach the rest of the world from China?

_____

_____

_____

_____

_____

Chapter 29 | **Soybean Protein**
Lesson Twenty-Nine

Introduction – Three Things You Know
- Write down three things you know about soybean protein.
- Share your knowledge with the class and make a list on the board.
- You will have assigned reading and will watch a Power Point presentation. Afterward, write down three things you've learned from both of these exercises. Save your answers to share later.

Reading
- Read the soybean protein chapter.

Power Point Presentation
- Watch the soybean protein chapter Power Point presentation.
- Write down three things you've learned. Remember to save your answers to share later.

Celebrity Profile
- Using the internet and other resources, choose an eco-minded celebrity to profile in a short, yet complete, one-page report. Be sure to include the reasons why the celebrity is eco-minded and list some of their associations to the sustainability community.

Three Things You've Learned
- Review your original soybean protein-related entries and compare them to what you've learned during the reading and Power Point exercises.

Soybean Protein Chapter Quiz
- Take the soybean protein chapter quiz.

Soybean Protein Chapter Quiz

Q1: What are the reasons for the skin enhancing properties of soybean protein?

_____

_____

_____

_____

_____

Q2: Why is soybean fiber also referred to as vegetable cashmere?

_____

_____

_____

_____

_____

Q3: Which famous American was among the first to use soybean fiber, for both his garments and automobile?

_____

_____

_____

_____

_____

Q4: Where was soybean protein fiber traditionally cultivated?

_____

_____

_____

_____

_____

Chapter 30 | Spider Silk
Lesson Thirty

Introduction – Three Things You Know
- Write down three things you know about spider silk.
- Share your knowledge with the class and make a list on the board.
- You will have assigned reading and will watch a Power Point presentation. Afterward, write down three things you've learned from both of these exercises. Save your answers to share later.

Reading
- Read the spider silk chapter.

Power Point Presentation
- Watch the spider silk chapter Power Point presentation.
- Write down three things you've learned. Remember to save your answers to share later.

Open Discussion
- What are the advantages of sustainable fabric production methods? Consider energy utilization, locality, transportation, local resources, labor, and costs. What are the disadvantages?

Three Things You've Learned
- Review your original spider silk-related entries and compare them to what you've learned during the reading and Power Point exercises.

Spider Silk Chapter Quiz
- Take the spider silk chapter quiz.

Spider Silk Chapter Quiz

Q1: What is another name for spider silk?

_____

_____

_____

_____

Q2: Why is spider silk called the *holy grail of fibers*?

_____

_____

_____

_____

_____

Q3: What quality of spider silk made it especially valuable for people in ancient times?

_____

_____

_____

_____

_____

Q4: Who pioneered the process of spider silk cultivation?

_____

_____

_____

_____

_____

Chapter 31 | Tencel® / Lyocell
Lesson Thirty-One

Introduction – Three Things You Know
- Write down three things you know about Tencel®/ lyocell.
- Share your knowledge with the class and make a list on the board.
- You will have assigned reading and will watch a Power Point presentation. Afterward, write down three things you've learned from both of these exercises. Save your answers to share later.

Reading
- Read the Tencel®/ lyocell chapter.

Power Point Presentation
- Watch the Tencel®/ lyocell chapter Power Point presentation.
- Write down three things you've learned. Remember to save your answers to share later.

My Own Way
- Have you ever been in a situation whereby you were confronted with challenges to your eco-friendly beliefs? What did you do and how did others react?

Three Things You've Learned
- Review your original Tencel®/ lyocell-related entries and compare them to what you've learned during the reading and Power Point exercises.

Tencel®/ Lyocell Chapter Quiz
- Take the Tencel®/ lyocell chapter quiz.

Tencel®/ Lyocell Chapter Quiz

Q1: What are some of the characteristics of Tencel® and how does it relate to lyocell?

_____

_____

_____

_____

Q2: Why do colors take on a rich look with Tencel®?

Q3: What quality does Tencel® impart to synthetic fibers?

Q4: Tencel® fiber is a sub-species of which synthetic fiber?

Chapter 32 | Wild Nettle
Lesson Thirty-Two

Introduction – Three Things You Know
- Write down three things you know about wild nettle.
- Share your knowledge with the class and make a list on the board.
- You will have assigned reading and will watch a Power Point presentation. Afterward, write down three things you've learned from both of these exercises. Save your answers to share later.

Reading
- Read the wild nettle chapter.

Power Point Presentation
- Watch the wild nettle chapter Power Point presentation.
- Write down three things you've learned. Remember to save your answers to share later.

News Article Review
- Research a current news article on eco-friendly fabrics. Describe its contents in detail, and provide insights to the class about its relevancy to sustainability.

Three Things You've Learned
- Review your original wild nettle-related entries and compare them to what you've learned during the reading and Power Point exercises.

Wild Nettle Chapter Quiz
- Take the wild nettle chapter quiz.

Wild Nettle Chapter Quiz

Q1: Which country is notably associated with wild nettle plant cultivation?

_____

_____

_____

_____

Q2: Which ancient tribe used wild nettle to make their fishing nets?

Q3: What is the connection between German soldiers and wild nettle?

Q4: What conditions are needed for the growth of the wild nettle plant?

Note Pages | **Student Workbook**
Guide to Green Fabrics

Guide to Green Fabrics Student Workbook™ — Chapter Lessons

Note Pages | **Student Workbook**
Guide to Green Fabrics

Note Pages | **Student Workbook**
Guide to Green Fabrics

Note Pages | **Student Workbook**
Guide to Green Fabrics

Note Pages | **Student Workbook**
Guide to Green Fabrics

Guide to Green Fabrics Student Workbook™ — Chapter Lessons

Note Pages | **Student Workbook**
Guide to Green Fabrics

Note Pages | **Student Workbook**
Guide to Green Fabrics

Note Pages | **Student Workbook**
Guide to Green Fabrics

Note Pages

Student Workbook
Guide to Green Fabrics

Note Pages | **Student Workbook**
Guide to Green Fabrics

Note Pages

Student Workbook
Guide to Green Fabrics

Guide to Green Fabrics Student Workbook™ — Chapter Lessons

Note Pages | **Student Workbook**
Guide to Green Fabrics

Guide to Green Fabrics Student Workbook™ — Chapter Lessons

Note Pages | **Student Workbook**
Guide to Green Fabrics

Guide to Green Fabrics Student Workbook™ — Chapter Lessons

Note Pages | **Student Workbook**
Guide to Green Fabrics

Note Pages

Student Workbook
Guide to Green Fabrics

www.ingramcontent.com/pod-product-compliance
Lightning Source LLC
Chambersburg PA
CBHW060518300426
44112CB00017B/2719